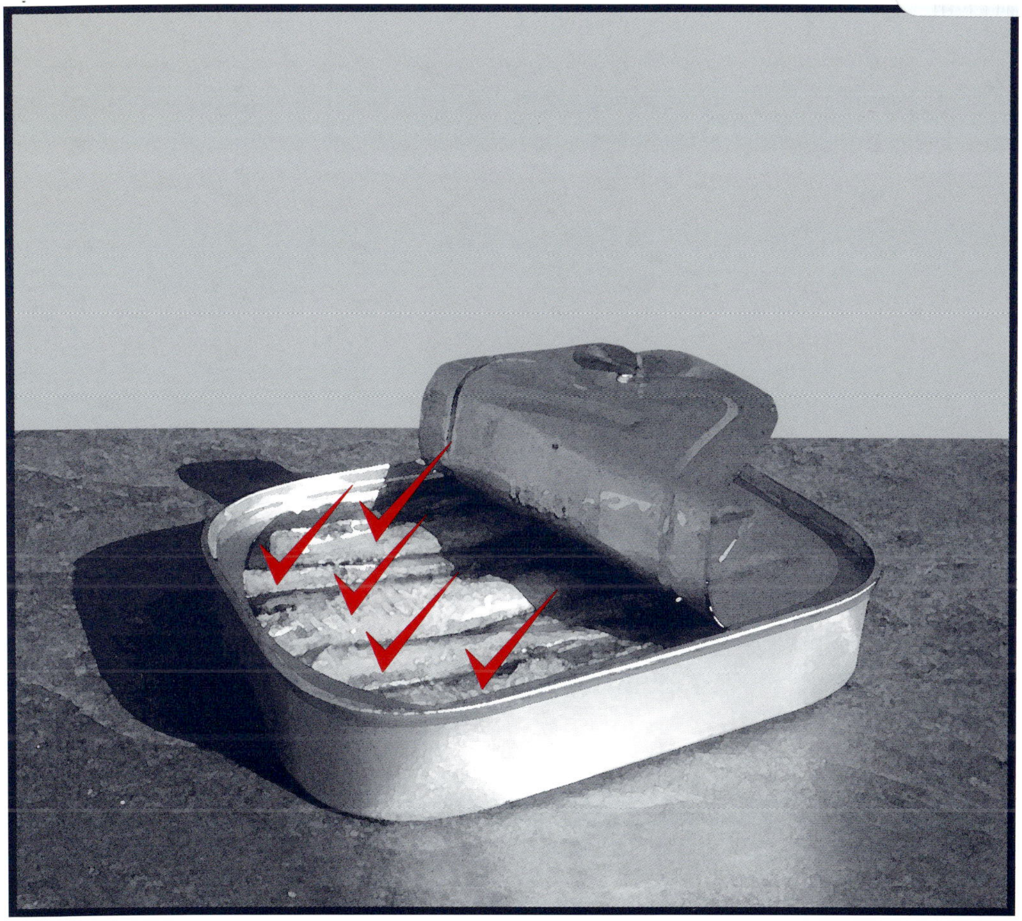

ECO-LABELLING AND SUSTAINABLE FISHERIES

Carolyn Deere

UCN
World Conservation Union

Food and Agriculture Organization of the United Nations

Citation: Deere, Carolyn L. (1999) Eco-labelling and Sustainable Fisheries, IUCN: Washington, D.C. and FAO: Rome.
ISBN: 28317-0507-X.
Available From: IUCN Publications Service
Unit 219c Huntingdon Road
Cambridge, CB3 ODL UK
Tel: +44 1223 277 894, Fax: +44 1223 277 175
Also available on IUCN website: http//www.iucn.org.
A catalogue of IUCN's publications is also available.

FAO Bookshop
United Nations Food and Agriculture Organisation (FAO)
Viale delle Terme di Caracalla
00100 Roma
Tel: ++39 06 5225 5688, Fax: ++39 06 5225 5155
Email: publications-sales@fao.org.
Also available on the FAO Fisheries Department website:
http://www.fao.org.

Author. Carolyn Deere wrote this paper as Policy Fellow for Trade and Biodiversity at the IUCN Washington Office. She is currently Warren Weaver Fellow at the Rockefeller Foundation. This document has benefited from input and comments from many sources. The author would like to thank David Downes, Jane Earley, Charlotte de Fontaubert, Scott Hajost, Sebastien Matthew, Jeff McNeely, Julia Novy, Jonathan Peacey, Sabrina Shaw, Mike Sutton, Despina Symons, Frank Vorhies, John Waugh and the staff of the FAO Fisheries Department for their comments and advice. She also thanks Amadou Tall of INFOPECHE and J. Santiago Caros Ros of INFOPESCA for the French and Spanish translations of the text. E-mail: cdeere@iucnus.org

TABLE OF CONTENTS

PREFACE

The need for sustainable fisheries and the conservation of marine biodiversity are both internationally recognised. Both IUCN and the FAO share the goals of supporting and strengthening international and national efforts to improve the management of fisheries, coastal areas and marine biodiversity. In principle, eco-labelling has been endorsed by the international community as one of the tools that can help improve environmental management through market-based means. However, its application to natural resource sectors has proven complicated and often controversial. The goal of this publication, jointly supported by IUCN and FAO, is to outline clearly some of the complications and concerns as well as the benefits in an objective fashion to provide an informed basis for moving forward.

ACKNOWLEDGEMENT

The author is Policy Fellow for Trade and Biodiversity, IUCN-Washington Office. She would like to thank David Downes, Jane Earley, Charlotte de Fontaubert, Scott Hajost, Sebastian Matthew, Jeff McNeely, Julia Novy, Jonathan Peacey, Sabrina Shaw, Mike Sutton, Despina Symons, Frank Vorhies, John Waugh and the staff of the FAO Fisheries Department for advice and comments.

INTRODUCTION

There is broad consensus in the international community that many of the world's commercial fisheries are in distress. Eco-labelling schemes are increasingly perceived as a way simultaneously to maintain the productivity and economic value of fisheries while providing incentives for improved fisheries management and the conservation of marine biodiversity. In the fisheries sector, a number of recent eco-labelling initiatives have been designed to complement and support efforts to implement sustainable fisheries management systems. These schemes have met with varying degrees of acceptance.

Aspects of eco-labelling that provoke concern include: the scientific basis of the criteria proposed for fisheries eco-labelling schemes; the attitude of consumers toward such schemes; and the potential impact of such schemes on international trade in fish and fishery products. In particular, there are concerns that eco-labelling schemes may cause discrimination against fish exports from developing countries and countries with economies in transition.

This short review is primarily for the consideration of policy makers, private industry and civil society (e.g. non-governmental organizations (NGOs), community-based organizations and fishers' organizations) in countries with a strong interest in eco-labelling of fish and fishery products. It is not intended to be a comprehensive overview of all aspects and technical considerations of eco-labelling. Rather, its goals are to clarify the key issues raised by eco-labelling and suggest the importance and benefits of stronger engagement by all countries and stakeholders in international discussions about different possible approaches to address them. The rationales for engagement include:

- There is a high probability that eco-labelling schemes for fisheries products are here to stay. Based on the experience in other sectors, one can expect that these initiatives will succeed in carving out a niche for their products and that rival schemes may develop. Even if some fail, there will likely be renewed efforts from governments, private industry and NGOs to initiate new ones.
- Countries have clear interests in safeguarding their trading interests and ensuring that eco-labelling schemes do not constitute disguised protectionism.
- Engagement offers developing countries, in particular, the opportunity to promote eco-labelling initiatives accompanied by adequate financial and technical resources in order to offset or compensate some of the costs they may incur in efforts to improve fisheries management and related implementation of international agreements.
- Processes such as the recent discussions in FAO fora provide an opportunity for all stakeholders to ensure that their varied interests are heard and promoted. By engaging in ongoing discussions countries can ensure that current and future eco-labelling schemes conform to basic principles regarding transparency and participation and that the criteria for eco-labels are developed with a view to their particular circumstances and needs.

This paper is divided into three parts:

- Part One reviews the current international framework and rationale for eco-labelling, the different types of eco-labelling schemes and examples of current eco-labelling initiatives in the fisheries sector and the relationship of eco-labelling to international trade rules.
- Part Two highlights the particular concerns and opportunities that eco-labelling schemes may present for developing countries.
- Part Three presents the case for stronger engagement in eco-labelling discussions by all stakeholders and examines the rationale for international technical guidelines.

I. INTERNATIONAL FRAMEWORK AND RATIONALE FOR ECO-LABELING

1.1. The International Framework for Eco-Labelling

There already is a common global understanding of the need for improved fisheries management and conservation of marine biodiversity. This follows from the 1982 UN Convention on the Law of the Sea and ensuing instruments, notably, the 1995 UN Agreement on the Conservation and Management of Straddling Fish Stocks and Highly Migratory Fish Stocks (Straddling Stocks Agreement), and the 1993 FAO Agreement to Promote Compliance with International Conservation and Management Measures by Fishing Vessels on the High Seas (Compliance Agreement). The 1995 FAO Code of Conduct for Responsible Fisheries and the technical guidelines developed in support of its implementation (such as those on the precautionary approach) provide further examples of international support for improved fisheries management. In addition, Agenda 21 of the UN Conference on Environment and Development (UNCED) held in Rio de Janeiro, Brazil and the 1992 Convention on Biological Diversity gave political support to the goals of improved fisheries management as well as to the conservation and sustainable use of marine biodiversity. Finally, the 1973 Convention on International Trade in Endangered Species of Wild Fauna and Flora (CITES) highlights international support for the principle of protecting endangered species.

The potential usefulness of eco-labelling schemes to create market-based incentives for environmentally friendly products and production processes was internationally recognised at UNCED. At Rio, governments agreed to "encourage expansion of environmental labelling and other environmentally related product information programmes designed to assist consumers to make informed choices."[1] Moreover, consumer organisations in many countries, and some international consumer unions, argue that consumers have a right to get information about products offered on the market that is relevant to their values and preferences, especially information pertaining to product safety or impacts on health or the environment.

Another basis for international eco-labelling efforts is also provided by the FAO Code of Conduct for Responsible Fisheries and other international and national instruments that emphasise the importance of achieving sustainability objectives through market-based measures and improving the identification of the origin of fish and fishery products traded (see Box I).

In recent years, there has been a proliferation of voluntary eco-labelling programmes for various products and sectors, many of which were initiated by NGOs and private industry as well as governments. All eco-labelling schemes share the common assumption that purchasing behaviour of consumers is not just motivated by price and mandatory quality and health standards. Rather, product attributes taken into account by consumers can relate to environmental and ecological objectives as well as economic and social objectives (e.g. fair trade; support to small farmers; discouragement of child labour).

[1]Paragraph 4.21 of Agenda 21.

Box I: Environment and Trade-related Provisions of Article 11 of the FAO Code of Conduct for Responsible Fisheries

- **Article 11.1.11.** States should ensure that international and domestic trade in fish and fishery products accords with sound conservation and management practices through improving the identification of the origin of fish and fishery products treated.
- **Article 11.1.12.** States should ensure that environmental effects of post-harvest activities are considered in the development of related laws, regulations and policies without creating any market distortions.
- **Article 11.2.3.** States should ensure that measures affecting international trade in fish and fishery products are transparent, based, when applicable, on scientific evidence, and are in accordance with internationally agreed rules.
- **Article 11.2.4.** Fish trade measures adopted by States to protect human or animal life or health, the interests of consumers or the environment, should not be discriminatory and should be in accordance with internationally agreed trade rules, in particular the principles, rights and obligations established in the Agreement on the Application of Sanitary and Phytosanitary Measures and the Agreement on Technical Barriers to Trade of the WTO.
- **Article 11.2.13.** States should cooperate to develop internationally acceptable rules or standards for trade in fish and fishery products in accordance with the principles, rights, and obligations established in the WTO Agreement.
- **Article 11.3.2.** States, in accordance with their national laws, should facilitate appropriate consultation with and participation of industry as well as environmental and consumer groups in the development and implementation of laws and regulations related to trade in fish and fishery products.

1.2. What are Eco-Labels?

Eco-labels are seals of approval given to products that are deemed to have fewer impacts on the environment than functionally or competitively similar products.[2] The rationale for basic labelling information at the point of sale is that it links fisheries products to their production process.

The goal of eco-labelling initiatives is to promote sustainably managed fisheries and highlight their products to consumers. Product claims associated with eco-labelling aim at tapping the growing public demand for environmentally preferable products. Eco-labels generally rely on life-cycle assessment to determine the environmental impact of a product 'from cradle to grave'.[3] Usually claims appearing on a product must be preceded by a chain of custody exercise that documents that the product was derived from, for example, a fishery certified as being 'sustainably managed'.

Prior to certification, a set of 'sustainability' standards or criteria against which a fishery is to be evaluated must be developed. Achieving and identifying 'sustainability' in fisheries is a complex process. The acceptance and credibility of standards is closely related to how the standards were developed, the standards themselves, and the accrediting or certifying process by which organisations are evaluated against the standard.[4]

[2]See OECD (1991) *Environmental Labelling in OECD Countries*, OECD Report 12, written by James Salzman, OECD: Paris and Karen West (1995) Ecolabels: The Industrialisation of Environmental Standards, *The Ecologist*, Volume 25, No. 1. See also Erika Preiss (1997) *An Ecolabel for Shrimp: Minimizing Potential Trade Barriers*, mimeograph prepared for International Environmental Law Clinic at NYU School of Law.
[3]See Elliot B. Staffin (1996) "Trade Barrier or Trade Boon? A Critical Evaluation of Environmental Labeling and its Role in the 'Greening' of World Trade", *Columbia Journal of International Environmental Law*, Volume 21, No. 2, p221.
[4]EDF (1997) *Global Deforestation, Timber and the Struggle for Sustainability: Making the Label Stick*, EDF: Washington, D.C.

ECO-LABELLING PROGRAMMES USUALLY FALL INTO ONE OF THE FOLLOWING CATEGORIES:

- First party labelling schemes: These are established by individual companies based on their own product standards. The standards might be based on criteria related to specific environmental issues known to informed consumers through the media or advertising. This form of eco-labelling can also be referred to as 'self-declaration'.
- Second party labelling schemes: These are established by industry associations for their members' products. The members elaborate certification criteria, sometimes by drawing upon external expertise from academia and environmental organisations. Verification of compliance is achieved through internal certification procedures within the industry, or employment of external certifying companies.
- Third party labelling schemes: These are usually established by a private initiator independent from the producers, distributors and sellers of the labelled products. Products supplied by organisations or resources that are certified are then labelled with information to the consumers that the product was produced in an 'environmentally friendly' fashion. The label (seal) is typically licensed to a producer and may appear on or accompany a product derived from a certified fishery or producer. Producers are usually expected to track the 'chain of custody' of their products in order to ensure that the products derived from the certified fishery are in fact those that are so labelled. In some instances the private initiator accredits other organisations to be the certifier. An accrediting body provides some degree of assurance that the certifier has been trained by an accredited training programme and is qualified to perform an evaluation against a specific set of criteria in a given field. While the criteria may be established through a negotiation process among the various interested parties, they are often motivated by the environmental objectives of the private initiators of such schemes. Environmental organisations and consumers generally prefer eco-labelling schemes of this type because of the heightened confidence that private commercial interests will not compromise the criteria applied to the schemes and strict compliance with them based on verifiable and impartial certification procedures.

Eco-labelling systems can be either mandatory or voluntary. Mandatory eco-labels are government-backed and could act as a trade restriction for foreign producers (i.e., imports may be rejected if they do not comply).[5] Imports of products that do not comply with voluntary eco-labels are not restricted. In the case of voluntary labels, it is up to the manufacturer to decide whether or not to apply for certification of the product, and the consumers choice whether to buy (or import) an eco-labelled product. Voluntary eco-labelling programmes may be funded and supervised by the private sector. Some, however, are government sponsored and funded.[6]

[5] WTO (1997b) *Eco-labelling: Overview of Current Work in Various International Fora*, Note by the WTO Committee on Trade and Environment, WT/CTE/W/45, WTO: Geneva; OECD (1997b) *Processes and Production Methods (PPMs): Conceptual Framework and Considerations on use of PPM-based Trade Measures*, OECD: Paris; OECD (1997c) *Eco-labelling: Actual Effects of Selected Programmes*, OECD, Paris.

[6] Germany became the first country with a government-sponsored eco-labelling programme when it began its Blue Angel label in 1977. The Blue Angel has appeared on products ranging from recyclable paper to detergents, vacuum cleaners and oil and gas heating appliances.

1.3. The Rationale for Eco-labelling in the Fisheries Sector

In the fisheries sector, there are hopes that eco-labelling schemes will:

- Provide information about the environmental impact of products and enable more informed purchasing behaviour by consumers and intermediaries;
- Provide consumers with the opportunity to express their environmental/ecological concerns through their purchasing behaviour and the market mechanism (e.g., dedicating their buying power to 'green catches'[7];
- Encourage retailers and consumers to buy only fishery products that come from sustainably managed resources;
- Raise environmental standards in the production of the commodity;
- Generate price differentials between eco-labelled products and those that either do not qualify for eco-labelling, or those whose producers do not seek to obtain such labelling[8];
- Enhance incentives for producers to supply products that meet the eco-labelling criteria in order to receive greater returns (a 'green premium') or gain market share for their products;
- Provide competitive advantages, market access or greater market share for fisheries products derived from sustainably managed fisheries; and
- Generate greater support by industry and other interested parties for improved fisheries management.

Eco-labelling schemes are often focused on domestic producers for the domestic market. Eco-labelling can also have the effect of enabling consumers to influence producers in other countries. A sizeable share—40 percent in 1996—of overall global fisheries production enters international trade.[9] This implies that eco-labelling has the potential to harness consumer preferences to create market-based incentives for sustainable fisheries management and improved production processes in other countries (such as harvesting methods that reduce by-catch, or fish caught in compliance with sustainable management regimes). Given that most trade in these products is destined for industrial country markets, eco-labelling schemes that focus on consumers in industrial country markets have the potential to encourage more sustainable international trade flows.

[7] Cathy Wessells of the University of Rhode Island has recently completed a consumer survey of U.S. seafood consumers, with a focus on determining if U.S. consumers have a preference for eco-labelled seafood, and are willing to pay for it. A report on the results of the survey can be found at http://www.riaes.org/resources/library.
[8] The FAO reports that for organic products a price premium of 10-20% is not difficult to obtain (and examples of premiums of as much as 50% have been reported). Premiums for certified forest products are estimated to be in the range of 5-10%. See FAO (1998) "Experience in eco-labelling of food and forest products", *Technical Consultation on the Feasibility of Developing Non-discrimination Technical Guidelines for Eco-Labelling of Products from Marine Capture Fisheries*, 21-23 October, 1998, FI:EMF/98/Inf.5, FAO: Rome, p. 10. A recent report by the MacArthur Foundation also discussed eco-labelling in the forestry sector and provided evidence of the price differential that eco-labels have created in that sector. See MacArthur Foundation (1999) *Sustaining Profits and Forests: The Business of Sustainable Forestry*, MacArthur Foundation: Chicago, p. 24-25.
[9] FAO (1999a) *The State of World Fisheries and Aquaculture*, FAO: Rome.

The success of certification and eco-labelling schemes as tools to encourage environment improvement is largely dependent on consumers' understanding and acceptance of certification, and in the general responsiveness of consumers to eco-labels. Whether the purchaser of the fisheries products is a major institutional buyer, a national or local government, or an individual consumer, the need for accurate and informative claims is equally great.[10] A 1997 report by the Environment Defense Fund highlighted that:

> "As certification programs proliferate, consumers and producers face choices as to which certification programs carry the most value. Consumers will need to understand who is doing the certifying, where the certification is credible, and to what standards the [fishery] has been certified. Differentiating between certification programs will be necessary unless a minimal set of criteria can be agreed upon internationally. The need to distinguish the accuracy and merit of competing certification claims may confuse consumers as well as producers and raise questions about which certification programs best serve the needs of the environment, the public and producers." [11]

There have also been concerns expressed about eco-labelling including that it could be used as a protectionist instrument in international trade and that it might discriminate against certain types of fisheries such as artisanal fisheries in developing countries which may lack the technical and financial resources to achieve certifiable standards.[12] These and other concerns are further discussed in Part II.

1.4. The Significance of Eco-Labels and the Potential for the Fisheries Sector

Voluntary eco-labelling schemes have become a fact for a wide range of products. The degree to which labels have captured market share varies depending on the product in question, and data concerning the market impact of eco-labelled products is very difficult to obtain. Similarly, hard data relating to the actual environmental impact of eco-labelling programmes is lacking.[13] In principle, the environmental impact of eco-labelling depends on the relevance and significance of the criteria as well as the market share of the eco-labelled products, which in turn depends on consumer preferences for eco-labelled products and the responsiveness of producers and suppliers.[14]

---

[10]EDF (1997) op. cit. (footnote 4) p45.
[11]EDF (1997) op. cit. (footnote 4) p45.
[12]On this point, see below footnote 82.
[13]See OECD (1997c) op, cit. (footnote 5). This paper focused on the following eight eco-labelling schemes: EU Eco-label Award Scheme, Swedish Environmental Choice, Nordic Swan, Canadian Environmental Choice Programme, Blue Angel, Green Seal, Japanese Eco Mark, and NF Environment. Most of these programmes are focused on products which reduce environmental damage during the use and disposal phase such as various types of detergents, cleaning agents and paper products. They encourage the use of recycled products and limit consumption of non-renewable resources. A limited number of eco-labels include requirements exclusively related to environmental effects which occur during the production phase (e.g. water effluents, air emissions). Few of the eco-labels in the selected schemes were developed for products of specific export interest to developing countries. The exception is the Nordic Swan programme which includes production related criteria which favour ecological cotton growing.
[14]See Preiss op. cit. (footnote 2) p.9; Rene Vossenaar (1997) *Eco-Labelling and International Trade: The Main Issues.* In Zarrilli, Simonetta, Jha, Veena & René Vossenaar (Eds.) (1997) *Eco-Labelling and International Trade, United Nations Conference on Trade and Development* (UNCTAD): New York.

In some markets (e.g., for household cleaning products) eco-labels have established a track record of promoting the spread of more environment-friendly production processes and product characteristics as well as raising consumer awareness about environmental issues.[15] So far, the results are more limited for natural resource based products such as organic and forestry products because eco-labelling schemes apply to only a very small share of production. Moreover, most schemes are too young to provide clear data. One exception is the single issue "dolphin safe" label attached to a large proportion of tuna products in the U.S. market. However, the label is ancillary to regulatory requirements, so labelling alone can not be identified as the primary cause of the high market share.

Nonetheless, there are strong indicators of the potential benefits to industries that do participate in eco-labelling schemes. The real significance of eco-labelling schemes stems not so much from presents sales or market share, but from the potential growth. Eco-labelling schemes in the fisheries sector also have significant potential due to the intense competition between retail chains of fishery products.

In the future, consumer consciousness of environmental concerns is likely to grow in both North and South. The market for eco-labelled organic products in several Europe and North America, for example, is said to be expanding more rapidly than supplies and average prices are significantly higher than for "non-organic" products.[16] There is evidence that credible eco-labelling schemes can induce a significant expansion of demand for tropical timber in eco-sensitive markets of Europe and North America.[17] In the fisheries sector, some retailers of fishery products have announced their intention only to buy certified fish in the future, and other large-scale wholesalers and retailers could follow suit. The potential impact for producers could be great as non-eco-labelled products could stand to lose real or relative market share.

Industry interest in eco-labelling stems, in part, from economic interests:

First, greening one's images is one of the primary strategies for product differentiation, profit, and market share in OECD markets. While there is no doubt that there will continue to be large markets (especially in Asia) for non-labelled fishery products, eco-labelled products could capture significant shares of most of the better-priced markets (e.g. in industrial countries) during the coming decade.

Second, some companies fear that growing public concerns about over-exploitation of marine fish stocks, environmental problems associated with fishery products and shrimp culture, as well as animal rights and health considerations (such as contamination) may spur a decline in demand for fish and fishery products. The adoption of eco-labelling schemes for fish and fishery products may be seen, to some extent, as an effort by industry associations and large-scale fish wholesalers and retailers to remove concerns among environment-conscious consumers about possible negative environmental impacts of their product choices and to retain market share and sustain demand for fishery products in countries where consumers are highly responsive to environmental issues (e.g., U.S., Germany, U.K., and Scandinavian countries). Eco-labelling is also one avenue that retailers and processors can use to ensure long-term supplies of fishery products that they market or process.[18]

---

[15]Drawing on case studies from the timber and organic foods sector, Kristin Dawkins (1996) provides substantial evidence that eco-labelling can be successful in meeting environmental objectives. She argues that, on balance, green products sell well and concludes that eco-labelling schemes enhance consumer education, and set minimum standards for environmentally-sound and socially just performance among other things. See Dawkins, Kristin (1996) "Eco-labelling: Consumer Right to Know or Restrictive Business Practice?" in Rüdiger Wolfrum (ed.) (1996) *Enforcing Environmental Standards: Economic Mechanisms as Viable Means?* Beiträge Zum Ausländischen Offentlichen Recht und Völkerrecht, Bd 125. Springer Verlag. Berlin. New York.

[16]See FAO (1998) "Experience in eco-labelling of food and forest products", *Technical Consultation on the Feasibility of Developing Non-discrimination Technical Guidelines for Eco-Labelling of Products from Marine Capture Fisheries*, 21-23 October, 1998, FI:EMF/98/Inf.5, FAO: Rome, p8.

[17]In the 1980s, concerned consumers, retail chains and institutional buyers including governments at local, provincial and national levels of several European and North American countries started to avoid, or even ban, the purchase of products made from tropical hardwood.

[18]Reduced supplies if coupled with lower demand could otherwise ultimately result in lower turn-over and reduced profits.

In sum, the possible benefits for industry from eco-labelling schemes include: a common understanding of preferred management practices and outcomes and thus reduced friction in international trade (assuming the schemes are internationally accepted); more efficient use of the fisheries resource base, fewer regulatory violations, greater market acceptance and improved public relations.[19]

1.5. Current and Proposed Eco-Labelling Initiatives for the Fisheries Sector

There are already several national, international, industry-sponsored, NGO-led and consumer-supplier partnership certification and standards schemes under development in the fisheries sector. The range of possible labels is broad. The focus of claims can range from "not over-fished, to no marine mammal by-catch and not over-fished, to no by-catch of any sort and not overfished, to ecosystem friendly where the entire ecosystem with its complicated food chain is not harmed".[20] The number of institutions and the diversity of their interests guarantee conflicts in the definition of what constitutes sustainable use of fisheries. This is all the more true since criteria for sustainability of fisheries are complex. Moreover, labels may be labelling entirely different things. For example, a standard indicating that a management system for sustainable fisheries is in place is not the same as certifying that a given consignment of fisheries products was sustainably produced, but both may appear on labels. The risk is that competing claims or conflicting labels will confuse consumers, causing them to lose confidence in certification and labeling and thus depriving the approach of its value.

Current eco-labelling initiatives relevant to the fisheries sector include first, second and third party eco-labelling schemes.[21] The following is a sample of prominent initiatives.

MARK OF ORIGIN

In many instances, producers have sought to gain competitive advantage by drawing attention to the origin of fish through labels. Moreover, the labelling of fish by origin and species is promoted by governments in some instances as a way to enable more effective tracking and identification of fisheries products to aid fisheries management (see Box II).

'DOLPHIN SAFE' LABELS

A variety of producers in the United States have made self-declarations that their tuna is 'dolphin safe'. The Dolphin Protection Consumer Information Act (DPCIA) of 1991 established criteria for the manner in which tuna must be caught. (On a voluntary basis, companies can then label their tuna to be 'dolphin safe'.)

ORGANIC SEAFOOD LABELS

There are also efforts underway by fishing companies in some parts of the world to label fish as farmed or wild, and more recently to win marketing niche with so-called 'organic seafood'. Organic labelling usually signifies that food has been produced without artificial inputs—especially synthetic fertilisers and pesticides—and has been grown using environmentally sound farm management techniques.[22] There are currently two pilot projects monitoring Alaska seafood

[19] EDF (1997) op. cit. (footnote 4).

[20] Cathy Wessells (1998) *Barriers to International Trade in Fisheries*, Discussion Paper prepared for the First FAO E-Mail Conference on Fish Trade and Food Security, October-November 1998.

[21] In addition to the listed initiatives, exploratory work is also undertaken at national and regional levels to examine the feasibility of eco-labelling schemes in fisheries (e.g. Nordic Council , see www. Norden.org.

[22] International Guidelines for the Preparation, Processing, Labelling and Marketing of Organically Produced Foods have been approved by the Codex Alimentarius Commission in July 1999. The Codex Alimentarius Commission is the body responsible for compiling the food standards, codes of practice, guidelines and recommendations that constitute the Codex Alimentarius; it operates under the auspices of the Food and Agriculture Organisation (FAO) and the World Health Organisation (WHO); see: http://www.fao.org/WAICENT/FAOINFO/ECONOMIC/ESN/codex

to help set standards to certify wild salmon as organic with the hope of breaking into the organic foods market.[23]

THE MARINE STEWARDSHIP COUNCIL (MSC)

The MSC is an independent, not for profit, international body headquartered in London, UK. It was initiated by the World Wide Fund for Nature (WWF) and Unilever, a large fish retailer, to promote sustainable and responsible fisheries and fishing practices worldwide. The MSC has, in collaboration with a selected group of parties interested in and experiences with fisheries issues, established a broad set of Principles and Criteria for Sustainable Fisheries.[24] Fisheries meeting these standards will be eligible for third party certification by independent certifying bodies accredited by the MSC. On a voluntary basis, fishing companies and organisations are expected to contact certifiers in order to have a certification procedure carried out.[25] Fish processing, wholesaling and retailing companies will be encouraged to make commitments to purchase fish from certified fisheries only. Unilever, for example, has pledged to buy only MSC certified fish by 2005. By opting to use the MSC logo, producers of fishery products are expected to give consumers the option to buy fishery products that have been derived from sustainable, well managed sources. Test cases for certification are presently being conducted.[26] The MSC offers stakeholders the opportunity to publicly endorse the organisation's mission, by signing a Letter of Support. More than one hundred fish retailing and wholesaling companies, fishing industry groups, fishers' organizations, certifying companies, conservation organizations, government ministries and others have signed such a letter.[27]

THE MARINE AQUARIUM COUNCIL (MAC)

MAC, a non-profit international organisation based in Hawaii (U.S.A.), brings together representatives of the aquarium industry, hobbyists, conservation organisations, government agencies and public aquariums. MAC aims at conserving coral reefs by creating standards and educating and certifying those engaged in the collection and care of ornamental marine life from reef to aquarium. It is working to establish standards for 'best practices' in the supply of marine aquarium organisms; an independent system to certify compliance with these standards; and consumer demand and confidence for certified organisms, practices and industry participants.[28]

---

[23]Proponents of organic labels for wild salmon argue that Alaskan salmon is intrinsically organic and that, provided it is free of prohibited additives throughout its life cycle, it should qualify as organic under the criteria for certification set out by the U.S. federal Organic Food Production Act of 1990. Already, some farmed salmon has been labelled organic because farmers could demonstrate a controlled environment and a diet consistent with the salmon's natural food. The organic food industry has been growing 20-24 percent annually over the last nine years compared to 3-5% growth of the conventional grocery industry. A difficulty that producers will face is proving that the fish have remained in a pristine environment while swimming through different waters. For further information see Dan Joling (1999) *Organic Seafood Cooking: State Backs efforts to Win Marketing Niche*, Associated Press, June 1999.

[24]According to the MSC, "A sustainable fishery is defined, for the purposes of MSC certification, as one that is conducted in such a way that: it can be continued indefinitely at a reasonable level; it maintains and seeks to maximise ecological health and abundance; it maintains the diversity, structure and function of the ecosystems on which it depends as well as the quality of its habitat, minimizing the adverse effects that it causes; it is managed and operated in a responsible manner, in conformity with local, national and international laws and regulations; it maintains present and future economic and social options and benefits; and it is conducted in a socially and economically fair and responsible manner". See www.msc.org.

[25]To date, the MSC has received eight applications from organisations to become accredited certifiers.

[26]Nineteen fisheries are currently candidates for MSC certification, and several test cases for fisheries certification are underway. These include the Western Australia Rock Lobster Fishery, the Thames Blackwater Herring Drift Net Fishery and the Dutch North Sea Herring Fishery. (www.msc.org)

[27]See www.msc.org. Information on the MSC's Principles and Criteria for Sustainable Fishing are available on that website including a list of companies and organizations that support the MSC's mission.

[28]http://www.aquariumcouncil.org/

THE RESPONSIBLE FISHERIES SOCIETY OF THE UNITED STATES (RFS) AND THE GLOBAL AQUACULTURE ALLIANCE (GAA)

RFS and GAA, headquartered in the U.S.A., have announced a joint eco-labelling scheme to recognize industry commitment and participation in responsible fisheries and aquaculture. The merger brings over 200 companies and individuals from 19 countries together in an effort to promote sustainable seafood harvest and production worldwide. The new eco-label will be offered to industry members who endorse the Principles for Responsible Fisheries of RFS or GAA's Principles for Responsible Aquaculture, and incorporate these Principles into their business. The RFS and GAA programmes are open to all segments of the industry (e.g., producer, importer, distributor, retailer or restaurant operator) and require the preparation of reports or plans that document implementation of the RFS/GAA principles. The RFS programme targets all types of domestic US seafood products while GAA focuses initially on farm-raised shrimp on a world-wide basis. GAA will conduct evaluations of shrimp farms based on a system of self-assessment questionnaires. The RFS is considering developing a third-party certification system.[29]

INTERNATIONAL ORGANIZATION FOR STANDARDIZATION (ISO)

General guidelines for environmental labels and declarations not addressed to any specific product category or sector are being developed by ISO, a non-governmental organization. ISO is a worldwide network of national standards institutes from some 130 countries, one from each country, with a central office in Geneva, Switzerland. For each country, the member body of ISO is the national body "most representative of standardization in its country". This national body may be from either the private or government sector. ISO's mission is to promote the development of standardization and related activities in the world with a view to facilitating the international exchange of goods and services, and to developing cooperation in the spheres of intellectual, scientific, technological and economic activity.

Environmental labels and declarations are one of the tools of environmental management, which is the subject of the ISO 14000 series.[30] This series does not prescribe environmental performance levels. Rather, to claim compliance with ISO 14000 standards, firms are required to establish an environmental policy and to set targets and objectives for environmental management performance.[31] ISO tends to be attractive to industry because it supports voluntary, market-based, measures as against traditional government command-and-control measures.

The ISO Sub-Committee on Environmental Labelling is responsible for developing standards in the field of environmental labels and declarations. The objective of the ISO 14 020 series is to set standards for the design and implementation of different types of eco-labelling programmes but not to lay down specific certification standards.[32]

---

[29]See www.nfi.org and http://www.gaalliance.org/GAA-RFSecolabel.html

[30]ISO 14000 is a series of international, voluntary environmental management standards. Developed under ISO Technical Committee 207, the 14000 series of standards address the following aspects of environmental management: Environmental Management Systems (EMS), Environmental Auditing & Related Investigations (EA&RI), Environmental Labels and Declarations (EL), Environmental Performance Evaluation (EPE), Life Cycle Assessment (LCA), and Terms and Definitions (T&D). (For further details, see http://www.tc207.org/faqs/index.html)

[31]Additional incentives to ISO 14000 implementation are: reduced environmental management costs due to the efficiencies of a systemic approach; potentially fewer regulatory violation and penalties since business would in theory better understand its environmental performance; improvement management of environmental risks and liabilities possibly leading to reduced insurance premiums; meeting customer demand; and improving public image.

[32]General principles for environmental labels and declarations were published in 1998 and standards for Type I eco-labels in April 1999. The ISO is developing standards for three different types of eco-labels. Type I eco-labels are those based on voluntary multi-criteria product life-cycle assessment of environmental effects;; verification is through a third party. Standards for Type II (self-declared environmental claims) and Type III (specialised third party schemes using quantified product information labels and pre-set indices) eco-labels are still under development. For further details see ISO (1998) *Environmental labels and declarations- General principles*, ISO 14020, Geneva; ISO (1999) *Environmental labels and declarations – Environmental labelling Type I – Guiding principles*. ISO 14024. Geneva.

BOX II: Labelling for Origin and Species

A preliminary step toward eco-labelling is simply the labelling of fisheries products according to origin and species. Identification of fisheries products by country of origin and species is not a counter proposal to certification or an alternative to it. Rather, it is an independent way to provide minimal information where none currently exists. The importance of the identification of the origin of fishery products was highlighted in the Code of Conduct for Responsible Fisheries. Article 11.1.11. of the Code states that "states should ensure that international and domestic trade in fish and fishery products accords with sound conservation and management practices through improving the identification of the origin of fish and fishery products treated". Furthermore, labelling for countries of origin and species would serve as a simpler and far cheaper test case for the working out of operational issues that bedevil certification schemes (e.g., accurate monitoring of chain of custody, products composed of multiple species, products processed in third countries).[33] It is also a first step toward building the consumer information and public consciousness needed to expand demand for eco-labelled products and to make ongoing eco-labelling efforts more effective.

The advantage of these simple identification-labelling efforts is that they need to wait for neither multilateral agreement on criteria for sustainability nor consensus on how best to apply them in order to get results.[34] Identification labels can also be used to track fisheries products and isolate those deemed to be caught illegally or in a fashion that undermines national or international management efforts. Moreover, labelling can be designed to be completely compatible with international trade regulations. The degree of compatibility, or conflict, with WTO rules depends on the specific form of labelling adopted, especially on the extent to which similar products are subject to equivalent requirements regardless of national origin.[35]

The WTO and the Mark of Origin: Article IX of the GATT explicitly accommodates national provisions for a Mark of Origin—such as a label identifying the country of origin—on imported products.[36] So long as the origin of like domestic products is equally identified, then the requirement for a mark of origin on all imported fisheries products should not constitute discrimination under WTO rules due to preferential treatment of national products.[37]

The WTO defines the point of origin as the "country where the good has been wholly obtained, or, when more than one country is concerned in the production of the good, the country where the last substantial transformation has been carried out".[38] For example, if a fish were caught in Namibia, but processed and packaged in Thailand, the country of last substantial transformation would be Thailand. This language restricts the information that countries can require on the ecological origin of imported processed fisheries products.[39] In order to permit explicitly the inclusion of information on the ecological origin (as opposed to the 'country of last substantial transformation'), WTO members would have to change the existing Mark of Origin language

As noted above, the identification of origin of fisheries products can provide a way to weed out those products deemed to be caught illegally or that are caught in a fashion that undermines national or international management efforts. For example, in recognition of the problem of trade in unreported, illegally harvested Patagonian toothfish, the Parties to the 1980 Convention on the Conservation of Antarctic Marine Living Resources (CCAMLR) have drafted a catch certification scheme for toothfish. The idea is that international trade in illegally caught Patagonian toothfish would be restricted by requiring that imports be accompanied by a valid certificate of origin.[40]

Similarly, the International Commission for the Conservation of Atlantic Tunas (ICCAT) introduced a Bluefin Tuna Statistical Document Programme for frozen bluefin (1992) and fresh bluefin (1993). The aim of the Programme was to increase the accuracy of bluefin statistics and track unreported fish caught by non-members and fleets flying flags of convenience. The Programme required all contracting parties to require that all imported bluefin tuna be accompanied by an ICCAT Bluefin Statistical document which details the name of the exporter and importer, the area of harvest, etc.[41] The Programme is equally applied to products from contracting or non-contracting parties to ICCAT. To ensure the effectiveness of the ICCAT bluefin tuna conservation program, the Commission has recommended in a resolution of 1995 that Contracting Parties take non-discriminatory trade restrictive measures, consistent with their international obligations, on bluefin tuna products in any form, from non-Contracting Parties whose vessels have been fishing for Atlantic bluefin tuna in a manner which diminishes the effectiveness of the relevant conservation recommendations.[42]

[33] EDF (1997) op. cit. (footnote 4).
[34] EDF, ibid.
[35] EDF, ibid
[36] WTO, GATT Article IX, p. 489.
[37] EDF (1997) op. cit. (footnote 4).
[38] WTO, Agreement on rules of Origin, Article 9.1. (b), 1995, p248.
[39] EDF (1997) op. cit. (footnote 4).
[40] CCAMLR (1998) Newsletter-December, CCAMLR: Hobart.
[41] WTO (1998) "Communication from the Secretariat of the International Commission for the Conservation of Atlantic Tunas (ICCAT)", Committee on Trade and Environment, WT/CTE/W/87, WTO: Geneva.
[42] See http://www.iccat.es/conserm.htm

1.6. The Relationship of Eco-labelling to International Trade Rules[43]

The issue of the interaction of eco-labelling schemes and international trade rules often confuses international discussions of eco-labelling questions. There appears to be a perception in some quarters that eco-labelling discussions at the World Trade Organisation (WTO) should be concluded prior to the development of international guidelines on this matter in other international fora.[44] However, it should be noted that the WTO does not claim to be the appropriate forum for discussions on the general usefulness of eco-labelling schemes or what constitutes appropriate criteria for assessing sustainability. Indeed, as the discussion below suggests, the WTO explicitly defers such issues to international agreements or bodies with appropriate expertise.

The objective of this section is to clarify what the Agreements of the WTO do and do not say about eco-labelling and to suggest that the respective roles of the WTO and other international organisations could be complementary with regard to eco-labelling.

1.6.1. The Agreement on Technical Barriers to Trade

The WTO Agreement that directly addresses eco-labelling is the Agreement on Technical Barriers to Trade (TBT).[45] WTO Members negotiated the TBT Agreement to ensure that members do not use technical regulations or standards as disguised measures to protect domestic industries from foreign competition. The TBT is also intended to reduce the extent to which technical regulations and standards operate as barriers to market access, primarily by encouraging the development of international standards. International standards are expected to reduce the obstacles to international trade that can be created by the proliferation of numerous different standards and regulations in various countries.

The TBT Agreement distinguishes between technical regulations and standards. `"Technical regulations" are defined as mandatory requirements for products or related process and production methods (PPMs). (Processes and production methods are defined as the way in which products are manufactured or processed and natural resources are extracted or harvested.[46]) "Standards", in contrast, are defined as voluntary requirements for products or related process and production methods.[47] Both regulations and standards may also relate to "terminology, symbols, packaging, marking or labelling requirements as they apply to a product, process or production method" (TBT Agreement, Annex I).

[43]The analysis of the TBT Agreement included in this section draws heavily from Downes, David R. (1999) *Integrating Implementation of the Convention on Biological Diversity and the Rules of the World Trade Organization*, IUCN Environmental Law and Policy Discussion Paper, IUCN: Cambridge.
[44]International organisation's currently engaged in eco-labelling include the Organisation for Economic Cooperation and Development (OECD), the International Organisation for Standardisation (ISO), the International Trade Centre (ITC), the United Nations Conference on Trade and Development (UNCTAD), and the Food and Agriculture Organisation of the United Nations (FAO). For an overview of recent work undertaken by several of these organisations see WTO Committee on Trade and Environment (1998) *Eco-packaging; overview of recent work in other International Fora*, WTO Doc WT/CTE/W/75, 5 March 1998, WTO: Geneva. A similar note, WT/CTE/W/45 (15 April 1997) was prepared by the WTO Secretariat on Eco-labelling. See www.wto.org.
[45]The WTO Secretariat notes that "well-designed eco-labelling programs can be effective instruments of environmental policy" so long as the key requirement of non-discrimination between foreign and domestic products is honored. See WTO Webpage on Eco-Labelling: www.wto.org/wto/environ/eco.html.
[46]OECD (1997c) op. cit. 137. p7.
[47]It is important to note that the TBT definition of standards differs from the definition of standards utilized by the ISO. Standards as defined by ISO may be mandatory or voluntary.

The rules of the TBT Agreement, including its Code of Good Practice for the Preparation, Adoption and Application of Standards (the Code of Good Practice), prohibit both regulations and standards from discriminating between domestic products and foreign products that are alike (the national treatment principle) and between 'like products' from different WTO Members (the 'most-favoured-nation' principle).[48] 'Like products' has been defined in past GATT and WTO dispute panel decisions to mean products with the same or similar physical characteristics or end uses. As a result, environmental trade measures based on distinctions between products based on their production or processing methods (PPMs) that do not in any way influence the physical characteristics of the products themselves have been found to violate these obligations (See Section 1.6.4.).[49]

The rules of the TBT also stipulate that Members shall ensure that technical regulations and standards do not create unnecessary obstacles to trade (TBT Article 2.2 and Annex 3). Furthermore, States are required to ensure that technical regulations use international standards that already exist (or that are near completion), or relevant parts of them, as a basis for their technical regulations, except when the international standards would be an ineffective or inappropriate mans for the fulfillment of the regulations objectives.[50] In the case of technical regulations, if a regulation is applied in accordance with a relevant international standard, it is presumed not to create an unnecessary obstacle to trade (TBT Article 2.5).[51]

International standards that could be recognised by the TBT include those set by central government, local government or non-governmental standardizing bodies.[52] International voluntary certification/labelling schemes and industry-led initiatives could possibly evolve to the point of serving as de facto international standards, without intervention from any inter-governmental process. The International Federation of Organic Agricultural Movements (IFOAM), a non-governmental body, for example, has established standards which are the basis for national organic labelling in several countries, and has publicly accepted the TBT's Code of Good Practice.

In terms of standards, Members must ensure that standardizing schemes operated by national governmental or intergovernmental agencies accept and comply with the Code of Good Practice (TBT Article 4.1). The extent to which the Code of Good Practice applies to local government and non-governmental standardizing bodies depends on them accepting and complying with it.[53] However, Members are required to take such reasonable measures as may be available to them to ensure that local government and non-governmental standardizing bodies as well as regional standardizing bodies accept and comply with the Code of Good Practice, irrespective of whether or not those standardizing bodies have accepted it(TBT Article 4.1).

The Code of Good Practice's substantive provisions require a standardizing body to, inter alia, 1) adopt existing or imminent international standards, except where they would be ineffective or inappropriate, 2) make reasonable efforts to harmonise standards at the international level, 3) make every effort to avoid duplication or overlap with the work of other standardizing bodies and achieve a national consensus on the standards they develop,[54] and 4) make

---

[48]The National Treatment Principle (Article III) forbids Members from treating foreign products less favourably (for example through more stringent regulation) than domestic "like products". The Most-Favoured Nation (MFN) principle (Article I) aims to prevent Members from treating products imported from one WTO Member less favourably than "like products" from another Member (Articles III and I).

[49]For more discussion of this point see Steve Charnovitz "Green Roots, Bad Pruning: GATT Rules and Their Application to Environmental Trade Measures, *Tulane Environmental Law Journal*, Vol. 7.

[50]For instance, because of fundamental climatic, geographical, technological and infrastructural factors; national security requirements; the prevention of deceptive practices; and protection of human health and safety, animal or plant life or health, or the environment. (TBT, Article 2.4. and 5.4.).

[51]On the question of whether a particular standard is in accordance with relevant international standards lies, the TBT does not indicate with whom the burden of proof lies. If a dispute did arise, there could be questions about: 1) whether a standard is in accordance with the relevant international standards; and 2) what constitutes a relevant international standard.

[52]Downes, David and Brennan Van Dyke (1998) *Fisheries Conservation and Trade Rules: Ensuring that Trade Law Promotes Sustainable Fisheries*, Center for International Environmental Law and Greenpeace: Washington, D.C., p.34.

[53]Appleton, Arthur, E. (1997) *Environmental Labelling Programmes: Trade Law Implications*, Kluwer Law International, p. 123-124.

[54]TBT Annex 3 does not specify precisely among whom the national consensus needs to be achieved. Presumably, the consensus should be among other relevant national standardizing bodies, but also with government, industry and NGOs (such as environmental and consumer organisations).

available to any interested party within the territory of a Member a copy of a draft standard submitted for comments, its most recent work programme and standards which it has produced (TBT Annex 3). While the TBT includes a specific statement that a technical regulation is applied in accordance with a relevant international standard is presumed not to create an unnecessary obstacle to trade (TBT Article 2.5), there is no similarly specific statement in the TBT or its Annexes on this issue with respect to standards.[55]

Finally, the TBT includes several specific provisions calling on all countries to ensure transparency in the development and application of standards and regulations in particular through the open dissemination of information about them.[56] It also calls on developed countries to recognize difficulties that developing countries may encounter in the formulation and application of technical regulations and standards, and to provide them advice and technical assistance for their endeavours in this regard (TBT, Article 11.). Developing country members are also to be provided differential and more favorable treatment given their special development, financial and trade needs (TBT, Article 12).[57]

1.6.2. The TBT Agreement and the Environment

The text of the GATT clearly states that some trade restrictions in the interest of conservation and animal and plant health are permissible, even though they violate the general principles of the GATT. The key Article of the GATT/WTO Agreements in terms of environmental issues is Article XX.[58] Article XX (b) of the GATT permits trade actions that are "necessary to protect human, animal or plant life or health". Article XX (g) provides for actions "relating to the conservation of exhaustible natural resources if such measures are made effective in conjunction with restrictions on domestic production or consumption". To qualify for any of these exceptions, a measure must also satisfy the requirements of the chapeau to Article XX.

While the TBT Agreement does not contain an explicit environmental exception, its preamble contains language paralleling that found in Article XX of the GATT. The preamble of the TBT Agreement recognises that "no country should be prevented from taking measures necessary to ensure . . . the protection of human, animal or plant life or health, [or] of the environment . . . at the levels it considers appropriate."[59] In addition, Article 2.2 of the TBT Agreement provides that the "legitimate objectives" of technical regulations include "protection of human health or safety, animal or plant life or health, or the environment."[60]

[55]On the question of whether a particular standard is in accordance with relevant international standards lies, the TBT does not indicate with whom the burden of proof lies. If a dispute did arise, there could be questions about: 1) whether a standard is in accordance with the relevant international standards; and 2) what constitutes a relevant international standard.

[56]This would include ensuring that an enquiry point exists which is able to answer all reasonable enquiries from other Members and interested parties and to provide documentation at an equitable price (if any) regarding adopted or proposed standards and technical regulations as well as conformity procedures (Article 10.1 and 10.4.). If a Member reaches agreement with another country or countries on issues related to technical regulations or standards which may have significant effects on trade, they are required to notify the Secretariat of the products covered by the agreement and provide a brief description of the Agreement (Article 10.7.).

[57]For example, in the preparation and application of standards and technical regulations, Members shall take account of the needs of developing countries with a view to ensuring they do not create unnecessary obstacles to the expansion and diversification of exports from developing country Members (Article 12). This may involve the provision of technical assistance, ensuring the active participation of developing country representatives in international standardizing bodies, and granting, upon request, specified, time-limited exceptions to obligations under the TBT.

[58]For a historical review of the world community's response to the link between trade and environment consult WTO Trade and Environment Division (1999) *High Level Symposium on Trade and Environment, Geneva 15-16 March 1999, Background Document*, and IISD Linkages: Trade & Sustainable Development, www.iisd.ca/linkages/trade.

[59]This language goes beyond the language of Article XX(b) in that it refers explicitly to the environment. Article XX(b) has, however, been interpreted by GATT and WTO panels to encompass measures generally considered environmental within its language referring to measures "necessary to protect human, animal or plant life or health."

[60]Also see Cameron, J. & H. Ward (1993) *The Uruguay Round's TBT Agreement*, WWF International: Gland; Vaughan, Scott & Ali Dehlavi (1998) *Policy Effectiveness & Multilateral Environmental Agreements*, UNEP Environment and Trade Series, No. 17. UNEP: Geneva; WTO (1995) *Negotiating History of the Coverage of the Agreement on TBTs with Regard to Labelling Requirements, Voluntary Standards, and PPMs Unrelated to Product Characteristics*, WT/CTE/W/10, WTO: Geneva.

1.6.3. The TBT Agreement and Eco-labelling

There is ongoing debate about how the TBT Agreement's different but related obligations on technical regulations and standards apply to eco-labelling initiatives. The WTO Secretariat suggests that the TBT agreement exerts "stronger control" over mandatory labels (those required by governments) than on voluntary or private eco-labelling schemes. However, the extent of control on each type of scheme is unclear.[61]

Eco-labelling schemes that are mandated by governments come clearly within the TBT's rules on technical regulations and other relevant WTO rules.[62] Voluntary, government and non-governmental labelling schemes also appear to be indirect targets of certain trade disciplines.[63] Members are required to take 'reasonable' measures to ensure that voluntary standardizing initiatives (which could include both government or non-governmental voluntary eco-labelling schemes) within its territory comply with the Code of Good Practice. (Analogous language found in the GATT requiring countries to take such 'reasonable' measures as are available to them has, in the past, been interpreted by dispute panels to require governments to take all constitutionally-available measures.[64])

Voluntary eco-labelling schemes for fisheries products do not appear, in principle, to contravene existing multilateral trade rules. The 1991 Tuna Dolphin decision of the GATT Arbitration Panel is instructive in this regard. While the panel found U.S. import restrictions adopted by the United States on tuna caught in association with dolphin to be GATT-illegal, it accepted the U.S. voluntary 'dolphin safe' tuna labelling scheme.[65] The panel noted that the voluntary label did not illegally restrict the sale of tuna since tuna products could be freely sold both with or without the 'dolphin safe' label, and because any competitive advantage conferred by the label depended on the free choice of consumers to give preference to tuna carrying the "Dolphin Safe" label.[66] While one could assume that a similar logic would apply to voluntary transnational eco-labelling schemes, to date, there is no similar precedent regarding the application of WTO rules to them.

1.6.4. The TBT Agreement and PPMs

Another unresolved issue is how the TBT Agreement applies to regulations or standards that invite consumers to discriminate not only on the basis of product characteristics, but according to PPMs.

Two kinds of PPMs with significant environmental impacts can be distinguished. First, a process or production method can affect the characteristics of a product so that the product itself may pollute or degrade the environment when it is consumed or used (product-related PPMs). Alternatively, a process or method itself can have negative impact on the environment through, for example, the manner in which natural resources are harvested or extracted in the production phase (non-product-related PPMs).[67] These production externalities do not affect the product characteristics.

[61]WTO (1998a) Regional Seminar on Trade and Environment for Developing Countries, Santiago, Chile, Trade and Environment Division, WTO: Geneva. p12
[62]Importantly, the TBT Committee of the WTO has decided that mandatory labelling requirements are subject to the notification provisions of Article 2.9 of the TBT Agreement, regardless of the kind of information that is presented. See G/TBT/1/Rev.3.
[63]Downes and Van Dyke (1998) op. cit.
[64]There is ongoing concern and debate about what the term all "constitutionally available" measures actually requires of governments.
[65]GATT, Dolphin-Tuna Panel, 1991, WL 771248 at *43.
[66]See GATT (1991) op. cit.
[67]WTO (1998a) op. cit.

Under WTO rules, the sovereign power of countries to restrict imports if they fail to meet domestic product regulations and standards relating to the physical characteristics of a product is left undisturbed. However, the power to make distinctions based on standards and regulations pertaining to PPMs which do not show up in the physical characteristics of the product is contested.[68]

Likewise the applicability of the provisions of the TBT Agreement to either mandatory or voluntary eco-labelling schemes that are based on non-product-related PPMs is also unclear, at best ambiguous and continues to be hotly debated.[69] Indeed, this has been an issue of much discussion in the WTO's Committee on Trade and Environment and Committee on Technical Barriers to Trade.[70] One issue on which there is broad agreement is that transparency plays a pivotal role in avoiding potential trade difficulties and increasing the legitimacy of such programmes and participation in them by parties interested in their development.

The interpretation of WTO rules with respect to the PPM issue is important because, in the context of eco-labelling, the most relevant regulations or standards are those relating to PPMs and their environmental impacts. Criteria for eco-labels for fisheries products are likely to be based on life-cycle analysis, whereby assessments of sustainability consider all phases of a product—production, processing, use and disposal. That is, eco-labels in the fisheries sector are likely to be predominantly awarded based on non-product-related criteria, particularly those related to harvesting methods (including type of gear used, level of by-catch, impacts on the marine habitats, compliance with management system and health of the stock of origin).[71]

Opposition to distinctions between products based on PPMs is often a strategy to guard against disguised protectionism. Within the CTE, there is recognition that standards related to non-product related PPMs will differ between countries due to a variety of factors. However, there are concerns that distinctions between products based on PPMs could be based on: a) arbitrary rationales that could undercut the principle of comparative advantage (for instance, regulations prohibiting products produced by workers earning less than a certain minimum wage); and b) well-intended but parochial understandings of what is environmentally sound that are derived from domestic ecological conditions which may not apply to conditions in distant countries. The prospect of distinctions based on PPMs also raises fears that some countries will be able to impose unfair economic pressure on other countries (frequently less developed than the importer) to match domestic environmental standards in their own jurisdiction or lose market access.[72] Developing countries, in particular, are often concerned that by broadening the scope of the GATT to permit distinctions based on environmental PPMs, they could be venturing toward a slippery slope whereby pressures for discrimination between products based on social PPM considerations (such as labour standards and human rights) might also intensify with even more significant potential trade ramifications.

Another argument presented against PPMs is that whereas conformity with product characteristic based standards can be assessed in either the producing country or the importing country, PPM-based requirements could be evaluated only on the site of production which could make this kind of assessment more expensive. Finally, there are concerns that PPM-based regulations might compel producers to use less efficient or costly technologies/methodolo-

[68]For example, the 1991 Tuna-Dolphin GATT dispute panel held that trade restrictions based on the process of creating a product, and not on specific qualities of the product qua product, are inconsistent with GATT. See Ted L. McDorman, "The 1991 US-Mexico GATT Panel Report on Tuna and Dolphin: Implications for Trade and Environment Conflicts", *North Carolina Journal of International Law and Commercial Regulation*, Vol 17., p461, p473.

[69]See, for example, CTE (1996) *Report (1996) of the Committee on Trade and Environment: Background, Analysis, Discussions and Proposals*, WT/CTE/1, WTO: Geneva.

[70]Ibid.

[71]Downes and Van Dyke, (1998) op. cit. p.1

[72]Downes, David R. (1999) *Integrating Implementation of the Convention on Biological Diversity and the Rules of the World Trade Organization*, IUCN Environmental *Law and Policy Discussion Paper*, IUCN: Cambridge.

[73]See footnote 54

gies, and/or restrict foreign suppliers' choice of technology.[73]

From a conservation perspective, the reluctance to permit PPM-based measures is problematic due to the increasing importance of PPM-based standards and regulations for effective environmental management. Domestic PPM-related measures are aimed at preventing environmental degradation caused by production processes, and as noted in a 1997 OECD Report, Domestic PPM-related requirements are important policy tools for promoting sustainable development.[74] Indeed, the conservation and sustainable use of fisheries depends on regulatory and management methods in the production phase (e.g., harvesting) as this is when considerable environmental impact occurs. PPM-related regulations and measures can be essential for controlling the environmental impact of consumption decisions. They also respond to the right of consumers to be informed about products they buy.[75] Finally, they offer the chance for greater efficiency because producers can compete to comply with standards in the most efficient way.

BOX III: History of Eco-Labelling Discussions in the WTO Committee on Trade and Environment (CTE)[76]

The Committee on Trade and Environment

When trade ministers approved the results of the Uruguay Round negotiation in Marrakech in April 1994, they also took a decision to begin a comprehensive work programme on trade and environment in the WTO. Trade ministers agreed to establish the WTO Committee on Trade and Environment (CTE) to cover all areas of the multilateral trading system—goods, services, and intellectual property.[77] The CTE has both analytical and prescriptive functions: "to identify the relationships between trade and environmental measures, in order to promote sustainable development; (b) to make appropriate recommendations on whether any modifications of the provisions of the multilateral trading system are required, compatible with the open, equitable and non-discriminatory nature of the system…".[78]

Two important parameters guide the CTE's work. First, WTO competence for work in this area is limited to trade and those trade-related aspects of environmental policies that may result in significant trade effects for its members. That is, there is no intention that the WTO should become an environmental agency, or that it should get involved in revising national environmental priorities, setting environmental standards or developing global policies on the environment. Second, if problems of policy coordination to protect the environment and promote sustainable development are identified through the CTE's work, steps taken to resolve them must uphold and safeguard the principles of the multilateral trading system.

[74]See OECD (1997b) op. cit. (footnote 5), p.7

[75]Ibid, p.3

[76]Sabrina Shaw in Duncan Brack (ed) (1997) *Trade and Environment: Conflict or Compatibility? Proceedings of the Royal Institute of International Affairs Conference*, Chatham House, London, April 1997, Earthscan : London.

[77]See WTO (1994) "Trade and Environment", Decision of 14 April, 1994, *WTO Document MTN/TNC/54* (MIN), WTO: Geneva.

[78]See WTO above footnote 52.

ECO-LABELLING IN THE CTE

Eco-labelling was extensively discussed in the GATT, and those discussions laid the basis in the CTE for a detailed examination of the issues involved.[79] The CTE's work programme covers 10 areas of which eco-labelling is one agenda item. Item 3(b) refers to the "relationship between the provision of the multilateral trading system and requirements for environmental purposes relating to products, including standards and technical regulations, packaging, labelling and recycling". The CTE's discussion to date has focused primarily on voluntary eco-labelling schemes and in particular on the transparency of such schemes.

The CTE adopted a report on 8 November 1996 for the first WTO Ministerial Conference in 1996.[80] That 1996 CTE report states that well-designed eco-labelling programmes can be effective instruments of environmental policy. It notes that in certain cases they have raised significant concerns about their possible trade effects. The report also noted that an important starting point for addressing some of those trade concerns is to ensure adequate transparency in their preparation, adoption and application, including affording opportunities for participation in their preparation by interest parties from other countries. The CTE stressed the importance of WTO Members following the provisions of the TBT Agreements and its Code of Good Practice. It also underlined the particular importance of ensuring fair access of foreign producers to eco-labelling schemes/programmes.

In 1997, the Committee on Trade and Environment produced a short overview of the work of various international fora in the field of eco-labelling.[81] In addition to outlining current work in other fora, it provides an overview of the work these institutions have completed, in particular in relation to the impact of eco-labelling schemes on market access.

---

[79] See GATT (1994) Report by Ambassador H. Ukawa (Japan), Chairman of the Group on Environmental Measures and International Trade, to the 49th Session of the Contracting Parties, WTO Document L/7402. February 2, 1994. GATT: Geneva.
[80] The Report WT/CTE/1 dated 12 November 1996, is available from the WTO Secretariat or http:www.wto.org.
[81] See WTO (1997b) op. cit. (footnote 5)

II. ECOLABELLING AND DEVELOPING COUNTRIES

Eco-labelling schemes, and in particular those that extend eco-labelling principles from household cleaning goods to agricultural and natural resource based products have provoked considerable concerns among some countries, particularly developing countries. To date, there is no conclusive evidence that eco-labelling schemes for other natural resources, such as forestry products, have, on average, proven detrimental to developing country interests. In terms of the fisheries sector, developing countries already have concerns about the impact on their competitiveness of rules related to fish additives and food safety, fish health and technical standards.[82] Their concern is that eco-labelling schemes in importing countries could simply add to the lair of constraints and competitive challenges they face. Several opportunities can also be articulated.[83]

2.1. Opportunities

Many industry groups, civil society organisations and governments acknowledge the economic and ecological opportunities that eco-labelling could offer.

ENVIRONMENTAL OPPORTUNITIES

Many governments and industry groups recognise that eco-labelling could provide needed economic incentives for better long term stewardship and availability of natural resources important for national economic welfare. Eco-labelling schemes can provide countries one tool to help them fulfill commitments made under international agreements on important environmental imperatives such as responsible fisheries and the conservation and sustainable use of biological diversity. The fundamental rationale for eco-labelling is, after all to generate political support for improved environmental management and to raise environmental standards through consumer choice.

ECONOMIC OPPORTUNITIES

Labelling provides one of the least-coercive market-based mechanisms to improve conservation outcomes.[84] Private sector interest in eco-labelling for fisheries products in both developed and developing countries is growing, especially given the business and export opportunities eco-labelling has generated in some other sectors. Moreover, as already noted, it is the potential for growth in the market share of eco-labelled products that makes eco-labelling a compelling business choice. If fisheries management improves in response to efforts to comply with certification criteria, the potential benefits to fisheries in both industrial and developing countries could go far beyond higher revenues which eco-labelled products may generate. In fisheries, there are clear win-win options, even if the task of fisheries management is daunting in many places.

Eco-labelling is seen by some as an important element for gaining access to new premium green markets (e.g., market access). For those producers willing and currently or potentially able to meet the sustainability requirements,

[82]Technical standards have been frequently used in the fisheries sector and have at times raised concerns about protectionist intents. There are strong fears that the introduction of Hazard Analysis Critical Control Point (HACCP) systems will represent potential non-tariff barriers to trade for some developing countries, especially in the case of non-modern production facilities. Fears that such measures can disguise protectionist intent led the members of the WTO to negotiate a series of agreements that regulate the use of non-tariff measures, including the Agreement on Sanitary and Phytosanitary Measures (SPS Agreement) and the Agreement on Technical Barriers to Trade.

[83]An excellent overview of the issues for developing countries is provided by Zarrilli et al (1997) op. cit. (footnote 12). The book brings together the papers presented by UNCTAD in June 1994 on possible effects of eco-labelling on export competitiveness and developing country firms' access to markets in developed countries.

[84]Ibid.

eco-labelling presents an opportunity to add value to existing products, expand reach in existing markets, or maintain market share in a competitive environment.[85] Product differentiation could be a way for some exporters to enhance their export earnings and eco-labels could be one source of such product differentiation.

There are also hopes that eco-labelling could provide new opportunities for attracting capital investment and joint ventures in developing countries. For example, some developing countries hope to enhance their chances at meeting criteria for the certification of their fisheries through cooperation among several countries in their region or through joint ventures with fishing enterprises from industrial countries. Eco-labelling can also provide an opportunity for innovative producers to benefit from the use of more environmentally-friendly production methods.[86]

Box IV: Support for Guidelines for Eco-labelling.

"...There are still interesting challenges in the area of trade and environment. One of these is the issue of eco-labelling.... Quite reasonably, some consumers are concerned as they approach fish counters and supermarket freezers and wondering whether their fish purchases are supporting similar disastrous exercises of overfishing. To the extent that customers are interested in being assured that the products they buy are harvested by sustainable fishing practices, and are prepared to pay more if necessary to buy products carrying the assurances they seek, we think they are entitled to reliable information in that direction. In this way, eco-labelling can harness consumer preferences through trade to strengthen sustainable fisheries management. For this reason, Namibia is supporting work by FAO towards the development of guidelines for eco-labelling."[87]

The Honourable Abraham Iyambo, Minister of Fisheries and Marine Resources, Namibia.

There are hopes that developing countries may be able to mobilise additional financial and technical resources through their participation in eco-labelling schemes. Conceivably, eco-labelling schemes could comprise specific support programmes to facilitate compliance by the private sector with the labelling criteria, especially in developing countries, as well as temporary measures to compensate individuals and households who may be negatively affected. Finally, some entrepreneurs in developing countries hope to carve out a distinct market niche based on the promotion of the sustainable nature of some artisanal modes of fish harvesting to both socially and environmentally conscious Northern consumers.[88]

In the future, consumer consciousness of environmental concerns is likely to grow in both North and South. This point is clearly recognised by many producers in both developed and developing countries. In both developed and developing countries, producers are working to comply with broad trends in environmental standards, such as ISO 14 000, in order to become more competitive in international markets.

In both North and South, one can argue that labelling that responds to consumer interest is likely to grow. Thus, at the global level, it makes sense for producers to get on board, one way or another, with environmental considerations in order to maximise their long term competitiveness. Moreover, it is notable that there are several producer organizations and NGOs in developing countries that recognise the opportunities that eco-labelling can present and that have had significant and productive involvement in the discussion of and development of eco-labelling schemes.

[85] See for example, UNCTAD (1994) *Eco-Labelling and Market Opportunities for Environmentally Friendly Products*, TD/B/WG.6/2. UNCTAD: Geneva.
[86] Downes and Van Dyke (1998) op. cit. p.33.
[87] Iyambo, Abraham (1999) "Fisheries, Trade and Environment: The Namibian Perspective", *Paper presented at the ICTSD-ZERO-ART Regional Trade and Environment Seminar for Government and Civil Society, Harare*, Zimbabwe, 10-12 February, 1999.
[88] Beatrice Chaytor (1999) "International Trade and Legal Rules to support Marine Biodiversity", *Fisheries, International Trade and Biodiversity*, draft manuscript, IUCN: Gland.

2.2. Concerns

Despite these opportunities, some governments, producers and civil society groups have expressed various concerns about eco-labelling.

First, an overriding complaint is of lack of transparency and opportunities for participation in the development of product standards such as those that might play a role in assessments of sustainability. This is of particular concern in the fisheries sector where governments have primary management responsibility for fisheries within national exclusive economic zones and, moreover, are obliged under international law to cooperate with governments of other countries in the management of shared fish stocks and of fish stocks on the high seas. Effective participation of governments in the product standard setting process may therefore contribute to strong implementation of eco-labelling programmes.

Second, there are concerns among some governments and industry groups, particularly those from countries with strong fish export interests, that eco-labelling schemes could a) disguise underlying intentions to protect domestic industries, b) restrict market access; and c) erode national competitiveness for those less able to meet or afford foreign labelling and certification standards.[89]

Possible discriminatory effects of national and regional eco-labelling schemes can be attributed to a number of factors, including: 1) eco-labelling tends to be based on domestic environmental priorities and technologies in the importing country and may overlook acceptable products and manufacturing processes in the country of production; 2) the definition of product categories, and the determination of criteria and limit values may favour domestic over foreign producers; 3) eco-labelling may require foreign producers to meet criteria which are not relevant in the country of production; 4) environmental infrastructures may differ widely across countries; and 5) certain parameters used for calculating the environmental effects of products throughout their life-cycle may be based on information collected in the importing country or countries with comparable conditions, and may overestimate the environmental impacts in the actual country of production.[90] Furthermore, given the influence of the voluntary purchasing decisions of large wholesale, retail and restaurant chains that control large market shares in large fish consuming and importing regions, particularly in Europe and North America, these schemes could effectively lead to reductions in the capacity of non-eco-labelled products to be exported to or simply sold within those markets.

Third, there are fears that the costs of bringing fisheries management practices into compliance with the criteria and principles of transnational or foreign eco-labelling schemes, going through the certification process, and maintaining certifiable status could be prohibitive.[91] One challenge is that the quantity and quality of fisheries data is often low in developing countries and this factor may be a constraint to certification.[92] Also, the burden of complying with foreign product standards may fall disproportionately on small suppliers to the market for whom the cost of acquiring

[89]See Downes and Van Dyke (1998) op. cit. p.145

[90]See Vossenaar, op. cit. (footnote 12).

[91]See Amjadi, Azita & Alexander Yeats (1995) *Nontariff Barriers Africa Faces: What did the Uruguay Round Accomplish, and What Remains to be Done?*, World Bank Research Working Paper 1439, World Bank: Washington, D.C; Gupta, R.K. (1997) *Non-Tariff Barriers or Disguised Protectionism*, Consumer Unity & Trust Society (CUTS): Calcutta; Matthew, S. (1997) *When Sandals Meet Suit: Letter from Sebastian Matthew, Executive Director of ICSF to Michael Sutton, Director, Endangered Seas Campaign, WWF International*, 7 August, 1997.

[92]Efforts are being made to address this problem by governments and through bilateral and multilateral assistance. The MSC has also stated its goal of ensuring that its Principles and Criteria can be applied in an appropriate manner in fisheries where there is limited information and where management and compliance regimes may be based on traditional community structures. Personal e-mail communication from Jonathan Peacey, Fisheries Director, MSC, October 1, 1999.

information about, and achieving, certifiable status and standards is relatively higher.[93] There have also been complaints that the lack of auditing/certification/eco-labelling infrastructure in developing countries will leave them dependent on expensive foreign consultants. As a result, developing countries have emphasised their need for greater financial and technical assistance for the improvement of fisheries management systems. The challenge of attaining sustainability is not unique to developing countries alone. Many fisheries in developed countries are depleted and unlikely to achieve certification in the near future. In developing countries, there are many fisheries that are less developed/depleted and for which certification might be more easily achieved. Therefore, in terms of the state of a fish stock, some certification programmes may in fact favour fisheries in developing countries over those in some developed countries.

Fourth, the voluntary nature of certification can raise challenges. While voluntary schemes need not result in explicit restrictions as some mandatory schemes might, they may indirectly affect trade due to institutional factors in producing countries. Institutional factors could include difficulties faced by producers in some countries in obtaining adequate supplies of materials, environmentally-friendly technologies and other materials which are acceptable for use in, or necessary to comply with standards for, eco-labelled products. Other institutional constraints could be inadequate and unequal financial and technical capacity within domestic regulatory agencies to facilitate sustainable fisheries management. Without the support of governments, many private industries can not reasonably be expected to become sufficiently organised to independently institute effective management schemes and achieve certifiable status. In cases where governments either fail to act (or act inappropriately) to manage fisheries, the fishing industry may be penalised due to lower sales prices in the absence of certification.[94]

Finally, it can be argued that even if participation in eco-labelling schemes is voluntary, the definition of criteria for certification could clearly influence the impact of the schemes on countries with varied environmental and socio-economic conditions and interests. In the absence of some common international understanding, governments could be required to try to monitor, intervene or improve each individual scheme that arises to ensure the interests of their countries are not compromised. International guidelines on eco-labelling could reduce this potential burden of monitoring. Otherwise, there is the possibility that promoters of voluntary competing eco-labelling schemes, for example at the national level, are likely to seek to discredit the schemes of competitors.

It is evident that the above concerns need to be addressed in one way or the other to make eco-labelling a widely acceptable, applicable and effective tool for attaining sustainable fisheries. In this regard, some avenues are outlined in part three of this publication.

---

[93] The WWF Endangered Seas Campaign and WWF US Marine Program have recently developed a proposed methodology for certification in community-based fisheries in part to address criticism that initiatives such as the MSC may disadvantage small-scale fishers from developing countries. They seek to generate 10 certified fisheries in marine eco-regions of broad geographical distribution in the next 3 years. Explicit goals are to test the potential of certification to create incentives for rationale resource exploitation and biodiversity conservation and to reward small-scale fishers for sustainable marine resource management. For more information see WWF (1999) *Community-Based Fisheries Certification; A Proposed Methodology*, WWF, Washington, D.C.

[94] It is possible that sufficient pressure from industry should induce governments to act. It is also possible, however, that industry has difficulty getting organised, and that government is unresponsive to industry pressure. Willmann, Rolf (1997) *Certification and Ecolabelling in Marine Fisheries: A Preliminary Assessment*, unpublished mimeograph.

III. THE CASE FOR STRONGER ENGAGEMENT

The simultaneous desire to ensure that eco-labelling produces positive environmental outcomes and is fair, and to protect industries from abuse, provides the case for greater international dialogue between and among governments, industry, environmental NGOs and civil society groups on eco-labelling in the fisheries sector.

Stronger international dialogue can serve several purposes. First, it could clarify the key issues around eco-labelling for policy-makers, industry, NGOs and civil society. In so-doing it may alleviate some of the concerns that lead some governments to discourage rather than support domestic industry and civil society groups that may stand to benefit from participating in eco-labelling initiatives.

Second, given that voluntary eco-labelling schemes are largely beyond the direct jurisdiction of the WTO, international dialogue is the strongest opportunity that governments have to promote eco-labelling schemes that: 1) make the strongest possible efforts to achieve the goal of improved fisheries management; 2) include credible fisheries management and environmental criteria; and 3) are designed to provide incentives and opportunities for a diverse range of producers and fisheries now or in the future.

Third, an international consensus on a set of technical guidelines both on the sustainability criteria for certification as well as on the certification process would help ensure that eco-labelling schemes do not become a mere marketing gimmick, but truly contribute to the ultimate goal of improving the sustainability of fisheries and conserving marine biodiversity. The existing text of the FAO's Code of Conduct for Responsible Fisheries would clearly be relevant to such guidelines, but additional aspects would also need to be elaborated such as, for example, procedural aspects of designing and running voluntary eco-labelling schemes, and guidelines for how equivalency of different eco-labelling schemes could be established (see Box V). A global dialogue appears to be especially needed on the following technical issues that have important implications for the environmental and economic impacts of eco-labelling schemes for fisheries products:

Box V: Equivalency and Mutual Recognition[95]

The concept of equivalencies entails the acceptance that when comparable environmental objectives can be achieved in different ways, different criteria should be accepted as a basis for awarding eco-labels depending on the environmental conditions of each country.[96] The concept of equivalency could be applied in several ways. For example, a voluntary or mandatory eco-labelling programme of an importing country might accept compliance with certain environmental requirements or the achievement of certain environmental improvements in the exporting country as 'equivalent' to compliance with the specific thresholds or criteria established in its programme. Alternatively, in the case of a mandatory scheme, the government of an importing country could decide that compliance with the exporting country's domestic regulations may be adequate for the award of an eco-label. Lastly, for intrinsically local problems in the producing country, a voluntary or mandatory eco-labelling programme in an importing country might accept as equivalent standards that are different but which remain helpful to the domestic environment of the producing country, taking into account its own environmental and developmental needs.

The most comprehensive concept of equivalencies could be mutual recognition, in which a product that qualifies for the eco-label of an exporting country programme is automatically awarded the eco-label of an importing country eco-labelling programme. Mutual recognition is most successful between countries with similar eco-labelling criteria". Some developing countries may be able to resolve concerns about potential cost-disadvantages in eco-labelling by promoting 'mutual recognition' by industrial countries of eco-labelling schemes initiated in their countries assuming they meet appropriate criteria.[97] One possibility would be to look for ways to integrate the International Accreditation Forum (IAF) regarding equivalence of accreditation programmes for certifiers in fisheries eco-labelling schemes. The IAF is an international body focused on improving the credibility of conformity assessments by developing transparency and accountability procedures for accreditation organisations that are responsible for granting certification. IAF membership includes accreditation bodies from nations in all parts of the world, industry representatives and accredited certification/registration bodies.[98]

[95]The information in this Box draws heavily from Erika Preiss (1997) op. cit. (footnote 2) and OECD (1997 b), op. cit. (footnote 5).
[96]Preiss (1997:16). See also George Richards (year) "Environmental Labelling of Consumer Products: The Need for International Harmonisation of Standards Governing Third-Party Certification Programs", *Georgetown International Environmental Law Review*, Vol.7, p 245, 248-261. New Zealand recently submitted a paper to the TBT Committee suggesting that equivalency of standards be developed to facilitate trade and avoid the difficulties that national standards can have on exporters. See G/TBT/W/88. The EC on the other hand, said that this was the most ambitious way to resolve problems in this regard.
[97]See UNCTAD (1994).
[98]Http://www.iaf.nu

- *The definition of a sustainable fishery.* In various international binding and voluntary instruments such as the 1982 United Nations Law of the Sea, the 1992 Convention on Biological Diversity and the 1995 Code of Conduct for Responsible Fisheries, States have given the undertaking to pursue certain objectives related to sustainability such as the conservation and management of fisheries resources and their habitats. There already exist several widely accepte generic definitions of sustainability.[99] These instruments and definitions provide the framework within which sustainability criteria for eco-labelling purposes in fisheries need to be laid down. However, issues may arise on how encompassing and detailed such criteria should be considering the bio-ecological, economic and social complexity and diversity of marine fisheries. One issue is, for example, whether a product's environment friendliness should be based solely on sustainability criteria of the fishery or fishery resources, or should take into account other environmental aspects such as fish habitats, eco-systems, or energy use in harvesting and processing. Another issue, that might be particularly important for developing countries, is whether criteria should also take into account economic, social and cultural criteria (such as impacts on fishworkers and local fishing communities). Some fishworkers organisations concerned about the impact of eco-labelling schemes on small-scale fisheries, particularly in the developing world, pro mote the concept of socially responsible fisheries whereby eco-labels would include considerations of local employment, working conditions and food security considerations.
- *Elaboration of global criteria that are also applicable to specific regions, countries and fisheries.* Diverse standards between different countries or regions can be warranted because: a) economic, social and environmental conditions differ from one country to another (what is appropriate in one set of circumstances may be inappropriate in another); b) national and regional weights given to conservation, economic, social and cultural sub-goals differ; and c) conditions in different fishery ecosystems and appropriate methods for management may differ depending on the country or situation. If eco-labelling standards take into account social factors, then the definition of what counts as sustainable will also vary according to economic, social and cultural factors. Eco-labelling systems that do not account or allow for such differences might discriminate against or work to the disadvantage of particular groups. The challenge is to strike a balance between this need for flexibility (e.g., specific standards for different regions, different types of ecosystems) and the need to ensure that there is some credible general principles or criteria that define what counts as sustainable fisheries management that is applied in as uniform a manner as possible. It is important to recall that the overall goal is to provide market-led incentives to raise standards and outcomes, not to find ways to legitimise current practices. On the other hand, given the proliferation of eco-labelling schemes, there is increased interest in harmonising the criteria for eco-labels and making the programmes more consistent. Harmonisation can be very useful for reducing the obstacles to international trade that can be created by the difficulty of complying with numerous, sometimes incompatible, standards and regulations in various countries.

[99]For example, the FAO definition reads as follows: "The management and conservation of the natural resource base, and the orientation of technological and institutional change in such a manner as to ensure the attainment of continued satisfaction of human needs for present and future generations. Such sustainable development conserves (land,) water, plants and (animal) genetic resources, is environmentally non-degrading, technologically appropriate, economically viable and socially acceptable"

- *The credibility of eco-labelled products.* Eco-labelled fishery products from both developed and developing countries face challenges concerning their credibility. In particular, the elaboration of different fishery or country-specific eco-labelling schemes with different standards and claims to fit the requirements of particular communities may jeopardise the credibility of labelling schemes in the eye of conservation-oriented consumers confused by different schemes. Indeed, confusion and concern could lead some consumers to withdraw from the market altogether.[100] Second, the task of assessing the sustainability of fisheries based on flexible criteria might result in high certification costs. This in turn might open the door to arbitrariness and/or the creation of loopholes in inspection and monitoring. Lack of transparency or inadequate expertise in eco-labelling programmes in developing countries could lead to doubts about the credibility of claims made by, and certified by, developing country accredited bodies.[101]
- *Process versus Performance Schemes:* An important debate regarding eco-labelling schemes is the usefulness of schemes that focus on ensuring that a management system or process is sufficiently focused on sustainability versus those that focus on the performance or outcome of the management system. Schemes that set standards for process or systems without prescribing sustainable outcomes, are not necessarily comparable to schemes that seek to grade performance and to ensure that products actually are produced in a sustainable way. A related issue is how the continued achievement of sustainable results can be maintained.
- *Technical and Financial Assistance Possibilities:* As noted above, an important concern that developing countries raise about eco-labelling schemes is that even voluntary eco-labelling schemes can have a discriminatory trade impact if the environmental criteria can only be met by countries that have certain financial and technological resources. Developing countries could use international dialogues as an opportunity to press for more action on the provision of technical and financial assistance to developing countries to comply with criteria for eco-labels. This could include swifter fulfilment of existing commitments made by the industrial countries to developing countries for financial and technical support to implement various international fisheries and environmental agreements. Developed countries and relevant international organisations could be encouraged to assist the switch to 'greener' production processes by: improving market research and disseminating this information to producers and trade associations; conducting workshops to inform producers about eco-labelling schemes; improving understanding and acceptance of environmental management schemes; improving capacity with regard to the technical skills needed to inspect and certify fisheries; and improving access to more sustainable fishing technologies and fisheries management equipment.[102]

---

[100] See Willmann, Rolf (1997) op. cit.; Tickell, Oliver (1999) "First the Forests, Now the Fish", *Green Futures*, May/June; Mattoo, Aaditya & Harsha Singh (1994) "Eco-labelling: Policy Considerations", *KYKLOS*, Vol. 47.

[101] If the prices of labelled and unlabelled products differ significantly, it may make the introduction of competing labelling schemes attractive. The challenge will be to ensure that certification procedures of these schemes verify that the source of the labelled products is indeed well-managed fisheries. In the timber industry, for example, eco-labelling of forest products proliferated in the 1980s. In a sample of 80 of such labels, a 1992 survey by the Worldwide Fund for Nature found that only three could offer any evidence to back up their claim of environment friendliness.

[102] Technical and financial assistance for sustainable development and environmental management is a priority for most, if not all, bilateral and multilateral assistance agencies (such as the FAO, the UN Development Program (UNDP) and the UN Conference on Trade and Development (UNCTAD)). Technical assistance to developing countries for trade and environment related issues, such as eco-labelling, is being provided by the WTO and UNCTAD and could be enhanced in the future. Some bilateral agencies and the World Bank are specifically investing in programmes to help small-scale producers qualify for certification under current eco-labelling schemes such as those promoted by FSC and MSC.

- *Trade-Related Issues:* It is both possible and useful to proceed simultaneously with dialogue at the WTO on particular trade-related issues of eco-labelling that demand attention, while engaging the FAO on issues of its specific competence such as the development of technical guidelines or criteria for judging the sustainability of fisheries, and the range of governmental and non-governmental organisations involved in developing generic guidelines on eco-labelling processes. Ideally, discussions in these fora would feed into each other. There are clearly a number of trade issues for eco-labelling that deserve further consideration and debate by both developed and developing countries. These include:
    - *The Applicability of the TBT Agreement:* In order to reduce uncertainty, the international community could consider developing a specific 'interpretation' of the TBT's applicability to both voluntary and mandatory eco-labelling schemes.
    - *PPMS:* As noted above the most-relevant category of PPMs for the fisheries sector is non-product-related PPMs. Further discussion is needed on how the use in eco-labelling programmes of criteria based on non-product-related process and production methods should be treated under the rules of the WTO Agreement on Technical Barriers to Trade. Several options for addressing non-product-related PPMs are conceivable (these need not be considered mutually exclusive).

First, future eco-labelling schemes could, in principle, ensure that their product-related or non-product related PPM-based standards are based on those already reflected in international agreements. The logic is that if non-product-related regulations or standards (such as eco-labels) can be shown to be consistent with standards included in an international agreement, they can be presumed not to create an illegal obstacle to trade. However, in practice, there are few international agreements that contain specific non-product-related PPM standards.[103]

A second approach thus would lie in consolidating support for the development of both product-related and non-product related PPM-based standards (or regulations) in international agreements. Moreover, international standards regarding PPMs are more likely to reflect legitimate differences in the available environment resources and their rate of depletion or in the acceptable level of risk.

Third, there could be more formal efforts to develop a process for formulating criteria for acceptable PPM-based measures and, in particular, eco-labelling schemes.

One international dialogue already underway to discuss and clarify some of these issues is a consultative process regarding the practicability and feasibility of developing global non-discriminatory technical guidelines for eco-labelling of products from marine capture fisheries hosted by the FAO.[104] At the first technical consultation in October 1998, a number of government delegations, especially from developing countries, were reluctant to engage fully in substantive discussion about these issues for several reasons including concern about promoting inadvertently the legitimacy, proliferation and/or expansion of eco-labelling efforts and a preference that the matter be dealt with under the auspices of the World Trade Organisation (WTO). The consultation, however, agreed upon some broad principles in the event that an agreement was eventually reached on the feasibility of elaborating technical guidelines for eco-labelling of fish and fishery products from marine capture fisheries (Box VI). More recently the FAO Fisheries Department has been given a mandate to continue its work on eco-labelling (See Box VII).

[103]The 1987 Montreal Protocol on Substances that Deplete the Ozone Layer is one multilateral environmental agreement that currently explicitly addresses the issue of both product and non-product related PPMs.

[104]For a detailed discussion of the issues surrounding the eco-labelling debate, and possible contents and format of technical guidelines on eco-labelling of products from marine capture fisheries see http://www.fao.org/fi/FAOCONS/FAOCE.asp.

Box VI: Paragraph 11 of the FAO Technical Consultation on the Feasibility of Developing Non-Discriminatory Technical Guidelines for Eco-Labelling of Products from Marine Capture Fisheries, Rome, Italy, 21-23 October 1998

There was a consensus that if an agreement was eventually reached on the feasibility of elaborating guidelines for eco-labelling this should be consistent with the Code of Conduct for Responsible Fisheries and, in addition, should consider inter alia the following principles:

- be of a voluntary nature and market driven;
- be transparent;
- be non-discriminatory, do not create obstacles to trade and allow for fair competition;
- establish clear accountability for the promoters of schemes and for the certifying bodies in conformity with international standards;
- there should be a reliable auditing and verification process;
- recognize the sovereign rights of States and comply with all relevant laws and regulations;
- ensure equivalence of standards between countries;
- be based on the best scientific evidence;
- be practical, viable and verifiable;
- ensure that labels communicate truthful information;
- must provide for clarity.[105]

One clear lesson from eco-labelling discussions thus far and in other sectors is that processes for developing labels and standards should not take place between governments alone, but should be conducted in fora that enable a broad range of stakeholders to be involved in their deliberations (e.g., relevant industry, environmental groups, development groups, scientists and civil society organizations with appropriate expertise on sustainability and fisheries).

[105]FAO. (1999) *Report of the Technical Consultation on the Feasibility of Developing Non-Discriminatory Technical Guidelines for Eco-Labelling of Products from Marine Capture Fisheries.* Rome, Italy, 21-23 October 1998. FAO Fisheries Report No. 594. Rome. 1999. 29 p. (http://www.fao.org/fi/FAOCONS/ecolab/R594e.asp)

BOX VII: Summary of FAO Discussions of Eco-labelling of Marine Fisheries Products

Relevant paragraph of the 116th Session of the FAO Council, Rome, Italy, 14 – 19 June 1999

- The Council recognised that eco-labelling was controversial and that FAO should continue its work in this area. The Council underscored the need to ensure that any eco-labelling scheme was transparent, voluntary, non-discriminatory and that it not be used to restrict trade. The Council recommended that FAO should contribute to the work of the World Trade Organisation's Committee on Trade and Environment and other bodies to help ensure that fish trade did not compromise responsible fisheries management.

Relevant paragraphs of the 23rd Session of the FAO Committee on Fisheries (COFI), Rome, Italy, 15 – 19 February 1999

- The Committee received the report of the Technical Consultation on the Feasibility of Developing Non-discriminatory Technical Guidelines for Eco-labelling of Products from Marine Capture Fisheries which was held in Rome from 21 to 23 October 1998, and expressed thanks to the Nordic Council of Ministers for providing the funding which made the Consultation possible.
- Various delegations stressed that the Consultation did not reach an agreement regarding the practicability and feasibility of FAO drafting technical guidelines for the eco-labelling of produce from marine capture fisheries. Having noted this, most delegations identified FAO as an appropriate international organisation to discuss technical criteria for eco-labelling schemes.
- It was agreed that the issue of eco-labelling could have significant implications for the fishery sectors of member nations.
- Most delegations endorsed the consensus expressed in paragraph 11 of the report of the Technical Consultation that if an agreement was eventually reached as to the feasibility of elaborating guidelines for eco-labelling, it should be consistent with the Code of Conduct for Responsible Fisheries, and cited the requirements of Articles 11.1, 11.2 and 11.3 of the Code of Conduct, which discuss responsible post-harvest utilisation, trade and regulations concerning fishery products. The Committee noted that any eco-labelling scheme must be transparent and not be an obstacle to trade, that it must be voluntary, non-discriminatory, ensure equivalence of standards between countries and/or schemes, recognise the sovereign rights of States, and comply with all relevant international agreements.
- Most delegations recommended that FAO should initiate further work on elaborating technical aspects of eco-labelling during the current and future biennia, subject to availability of resources. This work should include a review of what the WTO and other relevant international organisations are already doing.
- Other delegations, however, felt that FAO should not get involved in the issue and should leave it to the other specialist bodies.
- Some delegations also called upon FAO to convene a second open technical consultation on eco-labelling, In accordance with the rules of FAO governing such consultations and called upon the Secretariat to prepare relevant Terms of Reference in consultation with Member States.

Conclusion

There are clearly a number of issues regarding eco-labelling that deserve elaboration and discussion by the international community. It is vital that all governments as well as interested industry and civil society groups engage in these discussions to ensure that their interests are heard and appropriate responses developed and that proceedings for developing standards and systems for eco-labelling are transparent and advance the ultimate goal of ecologically responsible fisheries. Participation in the process of formulating sustainability criteria and certification processes is one way to ensure that the diversity of fisheries and interests in developing countries are considered.[106]

International efforts to promote a dialogue around the possible nature and content of global guidelines for eco-labelling provide an opportunity to develop an international understanding on appropriate certification criteria and processes. Endorsement of and participation in discussions, such as those hosted by the FAO, need not represent an endorsement by countries of existing or future eco-labelling schemes. By choosing to engage in such dialogue governments, industry, and civil society, conservation and fishworkers organizations can ensure that there is a broad framework in place for future and current eco-labelling efforts that can be used to: a) identify the particular concerns of developing countries; b) promote a set of parameters for eco-labelling schemes that are mutually beneficial for advancing strong environmental objectives and supporting the interests of industry.

[106]Matthew, S (1998a) "Preface", *Fish Stakes: The Pros and Cons of the Marine Stewardship Council Initiative: A Debate from the Pages of the SAMUDRA Report*, International Collective in Support of Fishworkers, Chennai: India, p1.

About IUCN.

Founded in 1948, the World Conservation Union brings together States, government agencies, and a diverse range of non-governmental organisations in a unique world partnership of over 880 members in all, spread across 133 countries. As a Union, IUCN seeks to influence, encourage and assist societies throughout the world to conserve the integrity and diversity of nature and to ensure that any use of natural resources is equitable and ecologically sustainable. A central secretariat co-ordinates the IUCN Programme and serves the Union membership, representing their views on the world stage and providing them with the strategies, scientific knowledge and technical support they need to achieve their goals. Through its six Commissions, IUCN draws together over 6,000 expert volunteers in project teams and action groups, focusing in particular on species and biodiversity conservation and the management of habitats and natural resources. Operations are increasingly decentralised and carried forward by an expanding network of regional and country offices, located principally in developing countries. IUCN builds on the strengths of its members, networks and partners to enhance their capacity and to support global alliances to safeguard natural resources at local, regional and global levels.

About FAO.

The Food and Agriculture Organization was founded in October 1945 with a mandate to raise levels of nutrition and standards of living, to improve agricultural productivity, and to better the condition of rural populations. Today, FAO is the largest autonomous agency within the United Nations system with 175 Member Nations plus the EU (Member Organization) and more than 4 300 staff members around the world. Following recent efforts to decentralize, FAO's staff includes almost 2 300 people at Headquarters and more than 2 000 working at decentralized offices and field projects. The Organization's 1998-1999 biennial budget is set at $650 million and FAO-assisted projects attract more than $3 000 million per year from donor agencies and governments for investment in agricultural and rural development projects. Since its inception, FAO has worked to alleviate poverty and hunger by promoting agricultural development, improved nutrition and the pursuit of food security. The Organization offers direct development assistance, collects, analyses and disseminates information, provides policy and planning advice to governments and acts as an international forum for debate on food and agriculture issues. FAO is active in land and water development, plant and animal production, forestry, fisheries, economic and social policy, investment, nutrition, food standards and commodities and trade. It also plays a major role in dealing with food and agricultural emergencies.

IUCN
The World Conservation Union

Food
and
Agriculture
Organization
of
the
United
Nations